Downton Abbey

Coloring Book

BuzzPop

BuzzPop
an imprint of Little Bee Books
New York, NY

For information about special discounts on bulk purchases,
please contact Little Bee Books at sales@littlebeebooks.com.
Manufactured in China RRD 0122
First Edition

10 9 8 7 6 5 4

ISBN 978-1-4998-0623-6

buzzpopbooks.com

Downton Abbey

Coloring Book

You'll find
there's never a dull moment in
this house.

Violet

I didn't run Downton
for THIRTY YEARS to see it go lock,
stock and barrel to a STRANGER from
GOD KNOWS WHERE.

Violet

Mary: I was ONLY going to marry him if NOTHING BETTER turned up.

Sybil: Mary, what a HORRID thing to say!

Mary: Don't worry. Edith would have taken him. . . .

I claim no career beyond the nurture of this house and the estate. It is my THIRD PARENT and FOURTH CHILD. Do I care about it? Yes. I DO care!

Robert

You'll SOON

get used to the way things

are done here.

Mary

Is there anything more THRILLING than a new frock?

Sybil

It's WORSE
than a shame. It's a
COMPLICATION.

O'Brien

You've shown me
I've been living in a DREAM, and it's
time to return to real life. Wish me
luck with it, Mary. GOD KNOWS
I wish the best for you.

Matthew

He's getting GRANDER than Lady Mary, and that's saying something.

Mrs. Hughes

What is the first law of service? We do not discuss the business of this house with STRANGERS!

Anna

I've told myself
and told myself that you're
too far above me, but
THINGS ARE CHANGING.

Branson

Richard: I'm LEAVING in the morning, Lady Grantham. I DOUBT we'll meet again.

Violet: Do you PROMISE?

You've lived your life
and I've lived mine. And now it's time we
lived them TOGETHER.

Matthew

Daisy: You've still kept me here with a DISHONEST REPRESENTATION.

Mrs. Patmore: Oh dear. Have you swallowed a DICTIONARY?

I've told you before,
if we're MAD enough to take on
the Crawley girls, we have to
stick together.

Matthew

You are being tested.
And do you know what they say,
my darling? Being tested only makes
you STRONGER.

Cora

There's nothing
MORE TIRING than WAITING
for something to happen.

Cora

Carson: HARD WORK and DILIGENCE weigh more than beauty in the real world.

Violet: If only that were TRUE.

Because you are my baby,
you know. And you always will be.
ALWAYS. My beauty and my baby.

Cora

Well, I have to
take one thing for granted.
That I will love you until the last breath
leaves my body.

Matthew

Revenge is SWEET.

Bates

Rose, you have OBVIOUSLY read too many novels about young women admired for their feistiness.

Edith

What with these

TOASTERS and MIXERS and such like,

we'd be out of a job.

Mrs. Patmore

SALT

When your only child dies,
then you're not a mother anymore.
You're not anything really.
And that's what I'm trying
to get used to.

Isobel Crawley

Life with me
won't be quite what
you're used to.

Michael Gregson

Honestly, Papa.
Edith's about as MYSTERIOUS
as a BUCKET.

Mary

Isobel: How you HATE to be wrong!

Violet: I wouldn't know. I am not familiar with the SENSATION.

A black singer
with the daughter of a marquess
in a North Yorkshire town.
Why should WE attract any
ATTENTION?

Jack Ross

Hope is a TEASE
designed to prevent us
accepting REALITY.

Violet

Why is it called
a WIRE-*LESS* when there's
so many WIRES?

Daisy

Oh, it is you.
I thought it was a MAN wearing
your clothes.

Violet

Do people think
we're some sort of hotel that
never presents a bill?

Robert

HA 9165

The presence of strangers
is our ONLY guarantee of
good behavior.

Violet

Well, the cat's away,
so we mice might as well
play a little.

Mrs. Patmore

Anna: Why are you smiling?

Bates: Because where I see a PROBLEM, you see only POSSIBILITIES.

BATSMAN
TOTAL
BATSMAN
No 1
0
LAST INNS
9266

I admit
it's an UNUSUAL sensation to
learn there's a secret in this house
I am ACTUALLY privy to.

Robert

Beware of being too good at it.
That's the danger of living alone.
It can be VERY HARD to give up.

Rosamund

Mrs. Hughes: You're not expecting a banquet, are you?

Carson: I'm expecting a delicious dinner prepared by the fair hands of my beautiful wife.

Mrs. Hughes: There's a THREAT in there somewhere.

There's no point

in even PRETENDING we can argue

with Lady Mary.

Robert

Edith: I SUPPOSE Cousin Isobel is entitled to put up an argument.

Violet: Of course she is. She is just not entitled to WIN it.

Anna: They do say OPPOSITES ATTRACT.

Mary: Yes, they attract. But do they live happily ever after?

I've done and said things. . . .
I don't know why, but I can't stop myself.
Now I'm paying the price.

Barrow

I believe
EDUCATION'S the gate
that leads to any FUTURE
worth having.

Molesley

MILTON

Violet: I do believe in rules and tradition and playing our part. But there is something ELSE.

Mary: And what is that, pray?

Violet: I believe in LOVE. Brilliant careers, rich lives, are seldom lived without some element of LOVE.

. . . In the end,
you're my SISTER and one day, only WE will remember Sybil or Mama or Papa or Matthew or Michael or Granny or Carson or any of the others who have peopled our youth, until at last our shared MEMORIES will mean more than our mutual dislike.

Edith

There they go,

a NEW COUPLE in a NEW WORLD. . . .

It seems all our ships are

coming into port.

Robert

It's strange,

but I feel completely,

completely HAPPY.

Edith